MERRILL
HEMISTRY
A GLENCOE PROGRAM

D0846266

dent Edition
cher Wraparound Edition
oratory Manual
oratory Manual Teacher Edition
nish/English Glossary
nputer Test Bank
eodisc Correlation
nsparency Package
blems and Solutions Manual
ving Problems in Chemistry
dy Guide, Student Edition

Teacher Resource Books:
  Enrichment
  Critical Thinking/Problem Solving
  Study Guide
  Lesson Plans
  Chemistry and Industry
  ChemActivities
  Vocabulary and Concept Review
  Transparency Masters
  Reteaching
  Applying Scientific Methods in Chemistry
  Evaluation

EVIEWERS

Vaine Alley
cie Southside High School
cie, IN 47302

nne Bowers
view High School
view, TX 79072

odore L. Boydston III
ace Curriculum Coordinator
e County Public Schools
ni, FL 33156

e Carr
ass High School
ago, IL 60628

tor James DeAlmeida
n High School
n, TX 77511

ia Shaw Dukes
Ley High School
Haven, FL 32444

Sandra Kay Enger
Nettleton High School
Jonesboro, AR 72401

Ronald R. Esman
Abilene High School
Abilene, TX 79603

William S. Frazer
Santa Teresa High School
San Jose, CA 95123

Janet P. Lee
Dr. Phillips High School
Orlando, FL 32819

Patricia Luco
Aldine Contemporary Education Center
Houston, TX 77088

Kathleen Munroe
Osceola High School
Kissimee, FL 34741

Erik Natti
Marblehead High School
Marblehead, MA 01945

Wylie C. Poulos
Cross Keys High School
Atlanta, GA 30319

James Eugene Sims
La Mirada High School
La Mirada, CA 90638

James F. Shannon
Pittsford-Mendon High School
Pittsford, NY 14534

Kelly Wedding
Santa Fe High School
Santa Fe, NM 87505

Earl Wilson
Cass Technical High School
Detroit, MI 49091

d all inquiries to:

:NCOE DIVISION
millan/McGraw-Hill
Eastwind Drive
terville, OH 43081

002-800803-0

ed in the United States of America.

5 6 7 8 9 0 - VH - 00 99 98 97 96 95 94 93

MERRILL
CHEMISTRY

AUTHORS
Robert C. Smoot
Richard G. Smith
Jack Price

Contributing Author
Tom Russo

GLENCOE
Macmillan/McGraw-Hill

New York, New York   Columbus, Ohio   Mission Hills, California   Peoria, Illinois

AUTHORS

Robert C. Smoot is a chemistry teacher and Rollins Fellow in Science at McDonogh School, McDonogh, Maryland. He has taught chemistry at the high school level for 33 years. He has also taught courses in physics, mathematics, engineering, oceanography, electronics, and astronomy. He earned his B.S. degree in Chemical Engineering from Pennsylvania State University and his M.A. in Teaching from the Johns Hopkins University. He is a Fellow of the American Institute of Chemists and a member of the National Science Teachers Association and the American Chemical Society.

Richard G. Smith is a chemistry teacher and Science Department Chairman at Bexley High School, Bexley, Ohio. He has been teaching chemistry at the high school level for 27 years. He received the outstanding teacher award from the American Chemical Society and has participated in NSF summer institutes in chemistry. Mr. Smith graduated Phi Beta Kappa with a B.S. degree in Education from Ohio University and earned his M.A.T. in Chemistry from Indiana University. He is a member of the American Chemical Society and the National Science Teachers Association.

Jack Price is co-director of the Center for Science and Mathematics Education at California State Polytechnic University, Pomona, California. He taught high school chemistry and mathematics in Detroit for 13 years, and then moved to San Diego to become the Math/Science Coordinator for the county. He earned his B.A. degree at Eastern Michigan University and M.Ed. and Ed.D. degrees at Wayne State University, where he carried out research on organometallic compounds. He has participated in NSF summer institutes at New Mexico State University and the University of Colorado. He is also an author of a high school mathematics textbook.

CONTRIBUTING AUTHOR and MICROCHEMISTRY SPECIALIST

Tom Russo teaches chemistry at the Millburn High School, Millburn, New Jersey. He is also the Science Supervisor for the Millburn Township school system. He received a B.A. in Science Education from Jersey City State College, an M.S. in biology from Seton Hall University, and an M.S. in chemistry from Simmons College in Boston. Mr. Russo has specialized in developing microchemistry methods for classrooms. He is a Woodrow Wilson, Dreyfus Master Teacher in Chemistry, and is the author of the laboratory manual that accompanies this program.

Content Consultants: Safety Consultant:

Teresa Anne McCowen
Chemistry Department
Butler University
Indianapolis, IN 46208

Mamie W. Moy
Chemistry Department
University of Houston
Houston, TX 77204

William M. Risen, Jr.
Chemistry Department
Brown University
Providence, RI 02912

Joanne Neal Bowers
Plainview High School
Plainview, TX 79072